Experiment No. -01

Theory

The Circuit in which resistances are connected end to end so that there is only one path for current to flow is a Series Circuit

1. The Total Resistance in the circuit is equal to the sum of individual resistances plus internal resistance of the cell if any

2. The Total Resistance is greater than the largest resistance in the circuit

3. The current I is the same in all parts of the circuit and hence the same reading is found on each of the two ammeters shown, and

4. The sum of the voltages V1, V2 ,V3,V4 is equal to the total applied voltage V ,i.e.

$$V = V1 + V2 + V3 + V4$$

Circuit Diagram

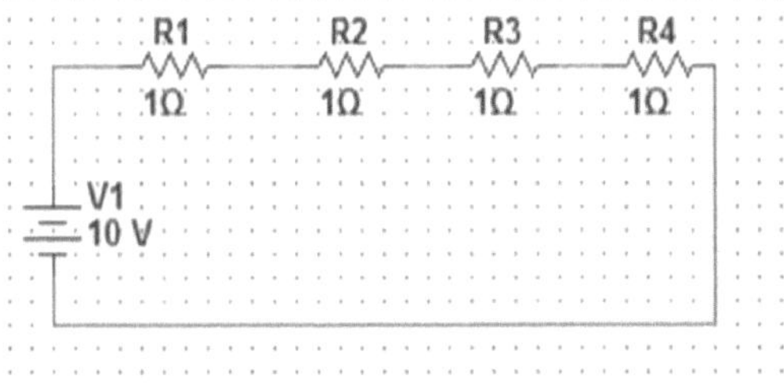

<u>**Code:**</u>

```matlab
rn=input('Enter the value of Resistance in row elements in a row     vector ');
v=input('Enter the value of DC voltage in volts');
req=sum(rn);
vn=rn.*v/req;
in=v/req;
pn=rn.*in^2;
i=v/req;
ptotal=v*i;
table=[rn' vn' pn'];
disp('');
disp('-------------------------------');
disp(' Resistance  Voltage  Power')
disp(' Ohms       Volts    Watts')
disp(table)
disp('-------------------------------')
fprintf('The Net Current is %f amps',i);
fprintf('\nPower Dissipated is %f watts',ptotal);
```

<u>OUTPUT(SIMULATION)</u>

```
Enter the value of Resistance in row elements in a row vector [1,1,1,1]
Enter the value of DC voltage in volts 10

-------------------------------

  Resistance  Voltage  Power
  Ohms        Volts    Watts
     1.0000    2.5000    6.2500
     1.0000    2.5000    6.2500
     1.0000    2.5000    6.2500
     1.0000    2.5000    6.2500


-------------------------------

The Net Current is 2.500000 amps
; Power Dissipated is 25.000000 watts>>
```

Experiment-02

% Aim: Plot a graph among

$$y_1 = x^2 + 2*(x+3) \text{ and } y_2 = 2*x^3 \text{ and } y_3 = x^4$$

Theory

A polynomial is a function of a single variable that can be expressed in the general form

$$A(s) = a1s\text{^}N + a2s\text{^ }(N-1) + a3s\text{^ }(N-2) + \cdots$$

+aNs+aN+1

Where the variable is s and the polynomial coefficients are theN+1 constants a1, a2, aN+1.The polynomial is of degree N, the largest value used as an exponent. The

Code

```
x=input('Enter the Range of Value of x= ');

y1=x.^2+2.*(x+3);

y2=2*x.^3;

y3=x.^4;

plot(x,y1,'r',x,y2,'b',x,y3,'o');

xlabel('value of x');

ylabel('value of y');

title('Graph of y1=x^2+2*(x+3) and y2=2*x^3and y3=x^4');

grid on;
```

OUTPUT

Enter the Range of Value of x= -10:1:10

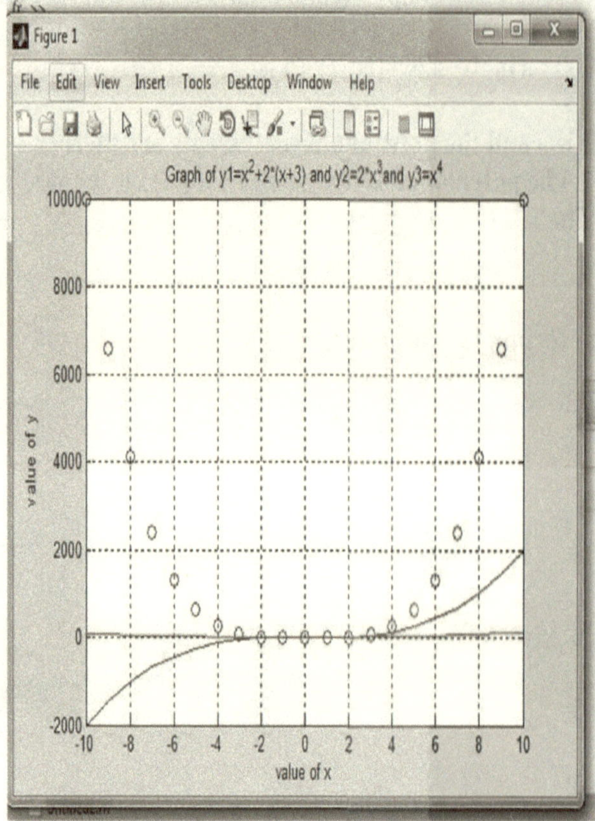

Experiment No.03

Theory

Ohm's Law

Ohm's law states that the current flowing in a circuit is directly proportional to the applied voltage V and inversely proportional to the resistance R, provided the temperature remains constant. Thus,

$$V=IR$$

Power Absorbed by A Resistor

A resistor, being a passive device, absorbs power. This absorbed power can be found from Ohm's law, that is,

$$V=IR \text{ and } P=VI=I^2R$$

The unit of power is the watt

1 watt = 1 volt × 1 ampere

Circuit Diagram

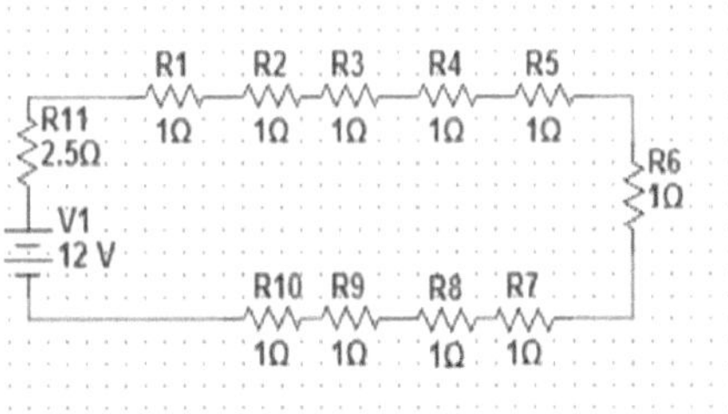

CODE

```matlab
vs=input('Enter the Source Voltage=');
rs=input('Enter Internal Resistance=');
rl=input('Enter the Range of Load Resistance=');

rtotal=rl+rs*ones(size(rl));
i=vs./rtotal;
pl=i.^2.*rl;
pmax=max(pl);
plot(rl,pl);
table=[i' rl' pl'];
disp('------------------------------------');
disp('----------START of TABLE----------');
disp('Current   RL      Power');
disp(' (Amps) (Ohms)   Watts');
disp(table);
disp('----------END OF TABLE------------');
xlabel('Load Resistances(in Ohms)');
ylabel('Power Delivered to Load(in Watts)');
axis('equal');
grid on
```

OUTPUT


```
Enter the Source Voltage=12
Enter Internal Resistance=2.5
Enter the Range of Load Resistance=1:1:10
------------------------------------
----------START of TABLE----------
Current       RL          Power
  (Amps)    (Ohms)        Watts
    3.4286    1.0000      11.7551
    2.6667    2.0000      14.2222
    2.1818    3.0000      14.2810
    1.8462    4.0000      13.6331
    1.6000    5.0000      12.8000
    1.4118    6.0000      11.9585
    1.2632    7.0000      11.1690
    1.1429    8.0000      10.4490
    1.0435    9.0000       9.7996
    0.9600   10.0000       9.2160

   ----------END OF TABLE--------------
```

Output Plot

Figure 1
File   Edit   View   Insert   Tools   Desktop   Window   Help
Power Delivered to Load(in Watts)
Load Resistances(in Ohms)

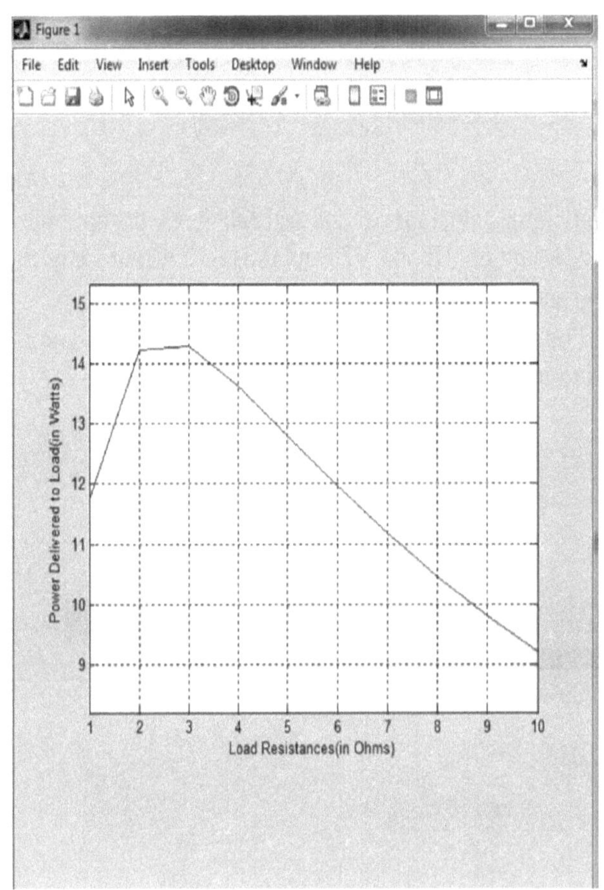

%Aim: Determination of Voltage Current and Power for %RL Circuit and Plot Graph

Theory

In an a.c. circuit containing inductance L and resistance R, the applied voltage V is the phasor sum of VR and VL, and thus the current I lags the applied voltage V by an angle ϕ lying between 0° and 90°

In any a.c. series circuit the current is common to each component and is thus taken as the reference phasor.

For the R–L circuit:

$$V = \sqrt{(V_R^2 + V_L^2)} \quad \text{(by Pythagoras' theorem)}$$

and $\tan \phi = \dfrac{V_L}{V_R}$ (by trigonometric ratios)

In an a.c. circuit, the ratio $\dfrac{\text{applied voltage } V}{\text{current } I}$ is called the **impedance Z,** i.e.

$$Z = \frac{V}{I} \ \Omega$$

The Phase Angle ϕ can be calculated using $\tan(\phi) = \dfrac{X_L}{R}$

If the Applied Voltage is $V = V_{max} \times \sin(\omega t)$, then the equation for the circuit current will be

$$I = I_{m} * \sin(\omega t - \phi)$$

Circuit Diagram

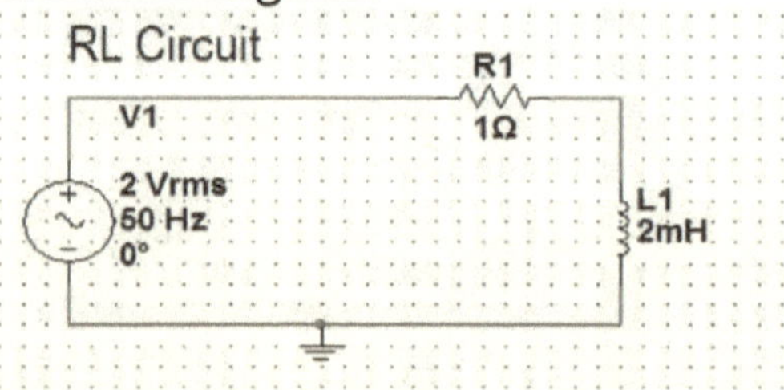

CODE

```
v=input('enter the value of source voltage in volts');
f=input('enter the source frequency in Hz');
r=input('enter the load resistance in ohm');
l=input('enter the load inductance in henry');
```

```matlab
t=input('enter the range of time in seconds');

vm=v*sqrt(2);
w=2*pi*f;
xl=w*l;
v=vm*sin(w*t);
phi=atan(xl/r);
wt=w*t;
z=sqrt(r^2+xl^2);
im=vm/z;
i=im*sin(wt-phi);
theta=w*t*180/phi;
p=v.*i;

plot(theta,v,theta,i,theta,p);
disp('------------------------------------');
disp('!!   voltage !!current !!power!!');
disp('!!   volts   !!amps   !!watts');
disp('------------------------------------');
table=[v' i' p'];
disp(table);
disp('------------------------------------')
disp('END');

xlabel('theta(wt) in degrees');
ylabel('voltage current power');
title('voltage,current,power vs wt');
grid on;
```

OUTPUT (INPUT SCREENSHOT)

```
enter the value of source voltage in volts2
enter the source frequency in Hz50
enter the load resistance in ohm1
enter the load inducatance in henry0.002
enter the range of time in secondslinspace(0,1/50,30)
----------------------------------------
!!   voltage !!current !!power!!
!!    volts     !!amps   !!watts
----------------------------------------
         0   -1.2741         0
    0.6080   -0.8084   -0.4915
    1.1876   -0.3049   -0.3621
    1.7117    0.2129    0.3644
    2.1557    0.7207    1.5536
    2.4989    1.1948    2.9858
    2.7253    1.6131    4.3962
    2.8243    1.9559    5.5240
    2.7912    2.2073    6.1609
    2.6275    2.3554    6.1890
    2.3411    2.3935    5.6033
    1.9451    2.3196    4.5118
    1.4582    2.1372    3.1165
    0.9031    1.8549    1.6752
    0.3058    1.4859    0.4544
   -0.3058    1.0474   -0.3203
   -0.9031    0.5599   -0.5057
   -1.4582    0.0463   -0.0675
   -1.9451   -0.4695    0.9133
   -2.3411   -0.9634    2.2554
   -2.6275   -1.4122    3.7107
   -2.7912   -1.7950    5.0102
   -2.8243   -2.0939    5.9137
   -2.7253   -2.2948    6.2541
```

```
   -2.4989   -2.3885    5.9686
   -2.1557   -2.3704    5.1100
   -1.7117   -2.2415    3.8368
   -1.1876   -2.0079    2.3846
   -0.6080   -1.6803    1.0217
   -0.0000   -1.2741    0.0000
```

OUTPUT WAVEFORM

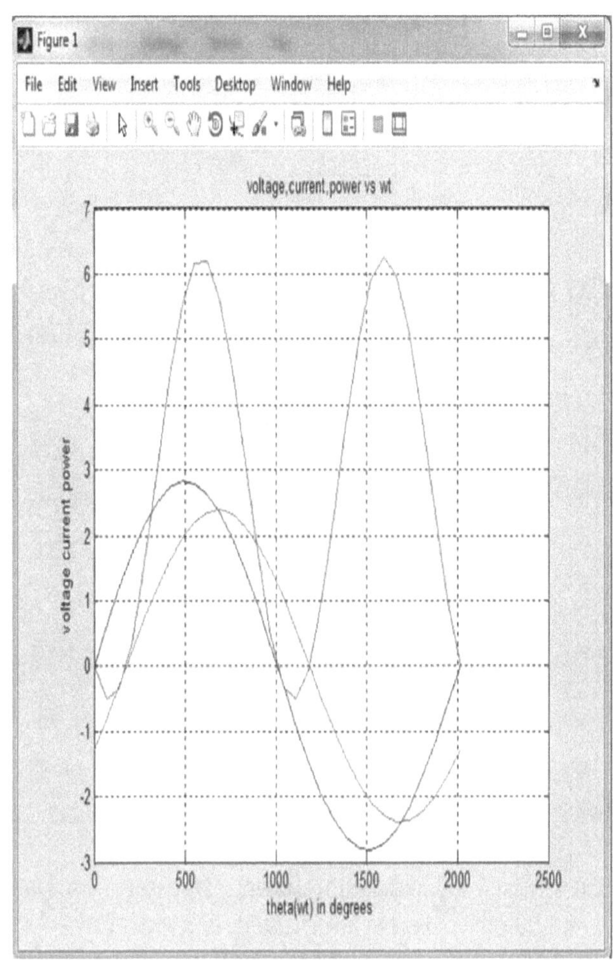

Experiment No.05

Theory

The maximum power transfer theorem states:

'The power transferred from a supply source to a load is at its maximum when the resistance of the load is equal to the internal resistance of the source.'

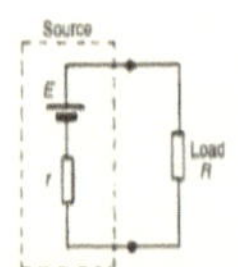

Here Maximum Power Transfer takes place in DC circuit When Load R=r

Load Resistance=Internal Resistance

v Efficiency of Maximum Power Transfer is only 50% as one half of the total power generated is dissipated in the internal resistance of the source

v Under the Conditions of Maximum power Transfer ,the Load Voltage is one-half of the open circuited voltage at the load terminals

v Maximum Power Transferred,$P_{max}=$

Typical practical applications of the maximum power transfer theorem are found in stereo amplifier design,seeking to maximize power delivered to speakers, and in electric vehicle design, seeking to maximize power
delivered to drive a motor.

Circuit Diagram with 1 Load Resistance

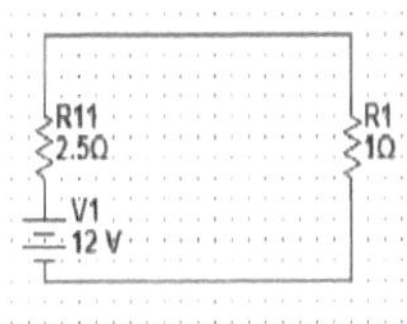

CODE

```
clc
vs=input('enter the voltage of dc source in volt=');
    rsint=input('enter the internal resistance of dc source in ohm');
rl=input('enter the range of load resistance in ohm');
rs=rsint*ones(size(rl));
rtot=rl+rs;
i=vs./rtot;
pl=i.^2.*rl;
plot(rl,pl);

xlabel('Load in (Ohms)');
ylabel('Power delivered to load in watts');
title('P vs RL curve');
table=[rl' rs' pl' ];
disp('--------------------------------');
disp('!!!RL   !!!Rs   !!PL !!');
disp('!!(Ohm)  !!(Ohm)  !!(Watts)');
```

```matlab
disp('--------------------------------');

disp(table);
disp('-------------End Of Table --------');
pmax=max(pl);
    fprintf('The Maximum power dissipated across load is %f watt\n',pmax);

grid on
```

OUTPUT

Command Window

File Edit Debug Desktop Window Help

New to MATLAB? Watch this Video, see Demos, or read Getting Started.

```
enter the voltage of dc source in volt=12
enter the internal resistance of dc source in ohm2.5
enter the range of load resistance in ohm1:0.5:10
--------------------------------
!!!RL   !!!Rs    !!PL !!
!!(Ohm)  !!(Ohm)  !!(Watts)
--------------------------------
    1.0000    2.5000   11.7551
    1.5000    2.5000   13.5000
    2.0000    2.5000   14.2222
    2.5000    2.5000   14.4000
    3.0000    2.5000   14.2810
    3.5000    2.5000   14.0000
    4.0000    2.5000   13.6331
    4.5000    2.5000   13.2245
    5.0000    2.5000   12.8000
    5.5000    2.5000   12.3750
    6.0000    2.5000   11.9585
    6.5000    2.5000   11.5556
    7.0000    2.5000   11.1690
    7.5000    2.5000   10.8000
    8.0000    2.5000   10.4490
    8.5000    2.5000   10.1157
    9.0000    2.5000    9.7996
    9.5000    2.5000    9.5000
   10.0000    2.5000    9.2160

-------------End Of Table --------
The Maximum power dissipated across load is 14.400000 watt
fx >>
```

OUTPUT PLOT

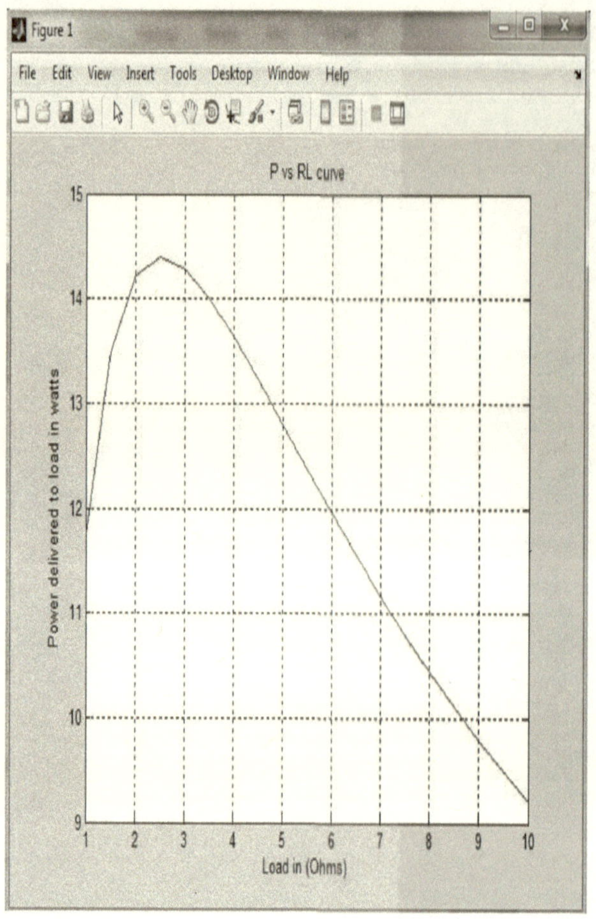

Figure 1
File   Edit   View   Insert   Tools   Desktop   Window   Help
P vs RL curve
Power delivered to load in watts
Load in (Ohms)

Experiment No.06

Theory

When an alternating voltage source is applied across pure resistance, then free electrons flow(i.e. current) in one direction for the first half cycle of the supply and then flow in the opposite direction in the next half cycle of the supply, thus constituting alternating current in the circuit.

The Alternating Voltage is $V = V_m \cdot Sin\omega t$

The Current in the Circuit $I = \dfrac{V_m}{R} \sin\omega t$

The Current and Voltage are in phase with each other

Power Dissipated across Resistor $= \dfrac{V_m}{\sqrt{2}} \cdot \dfrac{V_m}{\sqrt{2}}$

$P_{rd} = P_{av} = V_{rms}\, I_{rms} \cos\phi \quad \text{(in watt)}$

$Q = Reactive\ Power = V_{rms}\, I_{rms} \sin\phi \quad \text{(in VARs)}$

$P_a = Apparent\ Power = V_{rms}\, I_{rms} \quad \text{(in VA)}$

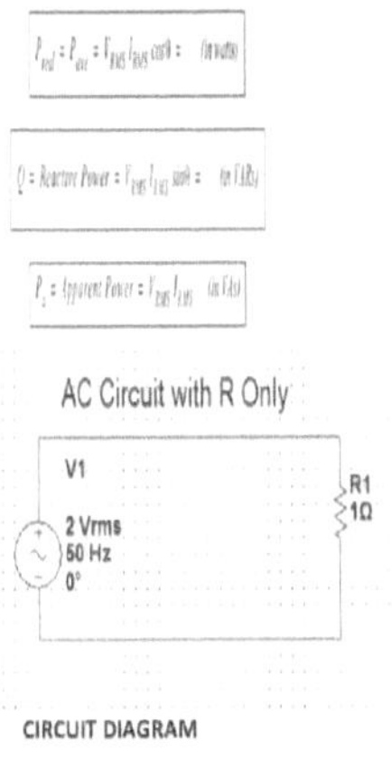

CIRCUIT DIAGRAM

CODE:

```matlab
v=input('enter the value of source voltage in volts');
f=input('enter the source frequency in Hz');
r=input('enter the load resistance in ohm');
t=input('enter the range of time in seconds');
vm=v*sqrt(2);
w=2*pi*f;
v=vm*sin(w*t);
theta=w*t*180/pi;
i=v/r;
p=v.*i;
plot(theta,v,theta,i,theta,p);
disp('------------------------------------');
disp('!! voltage   !!current  !!power!!');
disp('!! (volts)   !!(amps)  !!(watts)');
disp('------------------------------------');
table=[v' i' p'];
disp(table);
disp('------------------------------------')
disp('END');
xlabel('theta(wt) in degrees');
ylabel('voltage current power');
title('voltage,current,power vs wt');
grid on;
```

OUTPUT

```
Command Window
File  Edit  Debug  Desktop  Window  Help
New to MATLAB? Watch this Video, see Demos, or read Getting Started.
  enter the value of source voltage in volts2
  enter the source frequency in Hz50
  enter the load resistance in ohm2
  enter the range of time in secondslinspace(0,1/50,25)
  -------------------------------------
  !! voltage    !!current   !!power!!
  !!  (volts)     !!(amps)    !!(watts)
  -------------------------------------
          0          0          0
     0.7321     0.3660     0.2679
     1.4142     0.7071     1.0000
     2.0000     1.0000     2.0000
     2.4495     1.2247     3.0000
     2.7321     1.3660     3.7321
     2.8284     1.4142     4.0000
     2.7321     1.3660     3.7321
     2.4495     1.2247     3.0000
     2.0000     1.0000     2.0000
     1.4142     0.7071     1.0000
     0.7321     0.3660     0.2679
     0.0000     0.0000     0.0000
    -0.7321    -0.3660     0.2679
    -1.4142    -0.7071     1.0000
    -2.0000    -1.0000     2.0000
    -2.4495    -1.2247     3.0000
    -2.7321    -1.3660     3.7321
    -2.8284    -1.4142     4.0000
    -2.7321    -1.3660     3.7321
    -2.4495    -1.2247     3.0000
    -2.0000    -1.0000     2.0000
    -1.4142    -0.7071     1.0000
    -0.7321    -0.3660     0.2679
    -0.0000    -0.0000     0.0000
fx
```

OUTPUT WAVEFORM

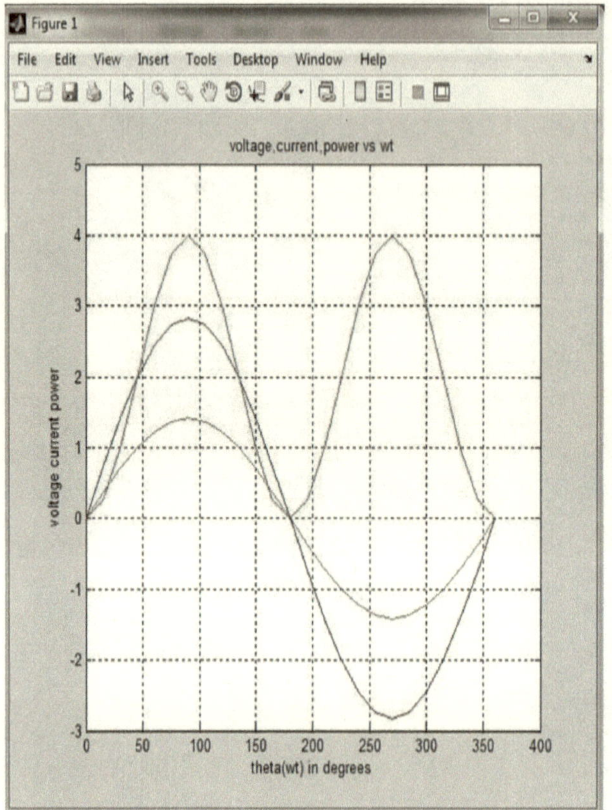

Figure 1
File   Edit   View   Insert   Tools   Desktop   Window   Help
voltage,current,power vs wt
voltage current power
theta(wt) in degrees

Experiment No.07

Aim: Determination of Current Resistance and Power for a Parallel Resistor Circuit and Plot Graphs of R vs I and R vs P

Theory

Two or more devices are said to be connected in parallel if and only if the same voltage exists across each of the devices.

- Ø When a number of resistances are connected in parallel, the reciprocal of total resistance is equal to the sum of the reciprocals of individual resistances

- Ø The Total Current in the Circuit is equal to the sum of currents in parallel branches
- Ø The Voltage across each resistor is same
- Ø The Total Resistance of the Circuit is always less than the smallest of resistances
- Ø The Conductances are Additive
- Ø The Powers are additive

$$\frac{1}{Rp} = \frac{1}{R1} + \frac{1}{R2} + \frac{1}{R3}$$

$$\frac{V^2}{Rp} = \frac{V^2}{R_1} + \frac{V^2}{R_2} + \frac{V^2}{R_3}$$

$$Pp = P1 + P2 + P3$$

Parallel Circuits are particularly useful in many electrical and electronic Circuits.

The Most Common Application Is a low resistor , called a shunt is connected in parallel with an ammeter to increase the current range of meter. If the Shunt is not used, the ammeter is able to measure current upto 1 A. Thus shunt increases the range of ammeter.

CODE

```
clc
```

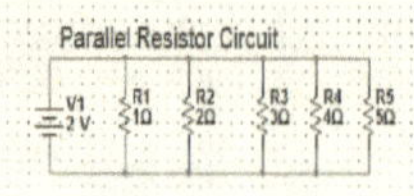


```
R=input('Enter the Resistance in row vector in Ohm');
V=input('Enter the Value of DC source');

N=1:1:max(size(R));
Req=1/sum(1./R);
I=V./R;
P=V*I;
Itot=sum(I);
Ptot=sum(P);
disp('------------------------------')
disp('------Start of Table-----------');
disp('!!Resistance !!Current !!Power');
disp('!!Ohms       !!Amps   !!Watts');
table=[R',I',P'];
disp(table);
disp('--------------End Of Table-------');
subplot(2,1,1);
plot(R,I);
xlabel('Resistance in Ohm')
ylabel('Current in Amps');
title('R vs I Plot');
grid on;

subplot(2,1,2);
plot(R,P);
```

```matlab
xlabel('Resistance in Ohms');
ylabel('Power dissipated in Watts');
title('P vs R Plot');
grid on;

fprintf('\nTotal Current in Amps is %f',Itot);
fprintf('\nTotal Power dissipated is %f watts',Ptot);
fprintf('\nEquivalent Resistance is %f',Req);
```

OUTPUT

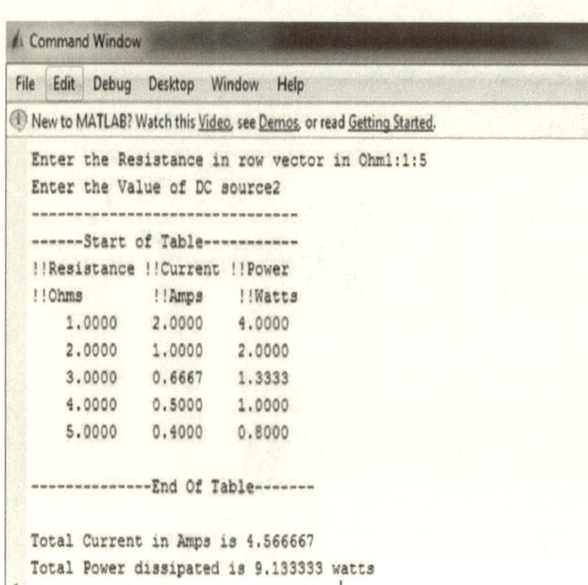

Command Window

File Edit Debug Desktop Window Help

New to MATLAB? Watch this Video, see Demos, or read Getting Started.

    Enter the Resistance in row vector in Ohm1:1:5
    Enter the Value of DC source2

    ------Start of Table-----------
    !!Resistance !!Current !!Power
    !!Ohms   !!Amps   !!Watts
         1.0000   2.0000   4.0000
         2.0000   1.0000   2.0000
         3.0000   0.6667   1.3333
         4.0000   0.5000   1.0000
         5.0000   0.4000   0.8000

    --------------End Of Table-------

    Total Current in Amps is 4.566667
    Total Power dissipated is 9.133333 watts
    Equivalent Resistance is 0.437956>>

Subplots

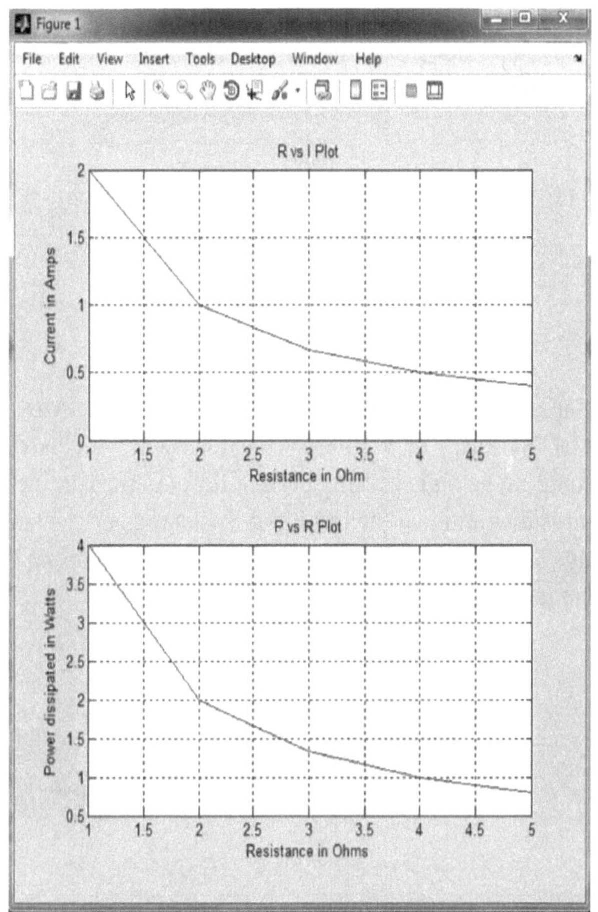
Figure 1
File   Edit   View   Insert   Tools   Desktop   Window   Help
R vs I Plot
2
1.5
1
0.5
0
Current in Amps
1   1.5   2   2.5   3   3.5   4   4.5   5
Resistance in Ohm
P vs R Plot
4
3.5
3
2.5
2
1.5
1
0.5
Power dissipated in Watts
1   1.5   2   2.5   3   3.5   4   4.5   5
Resistance in Ohms

Experiment No.08

Aim%To Determine the Charging Discharging current for a capacitor and Plot the graphs of current vs time

Theory

When an uncharged capacitor of Capacitance C farad connected in series with a resistor R to a DC SUPPLY of V Volts. When the Switch is closed, the Capacitor starts charging up and charging current flows in the Circuit .The Charging Current is maximum at the instant of Switching and decreases gradually as the voltage across the capacitor is charged to applied voltage V,the Charging current becomes zero.

Using KVL,

CIRCUIT DIAGRAM

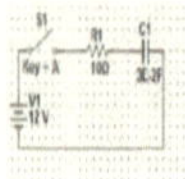

$$V = V_R + V_C$$

$$V = CR\frac{dV_c}{dt} + V_c$$

On Simplifying Linear Differential Equation

(Variable Separable)

Voltage across Capacitor,

$$V_C = V\left(1 - e^{-\frac{t}{RC}}\right)$$

Charge on Capacitor

$$q = Q\left(1 - e^{-\frac{t}{RC}}\right)$$

Charging Current, $I = \frac{V}{R}e^{-\frac{t}{RC}}$, $\tau = RC$, is the Time Constant

For Discharging

$$V_C = V e^{-\frac{t}{RC}}$$

$$I = \frac{V}{R}e^{-\frac{t}{RC}}$$

-ve Sign is attached because the Discharging Current flows in the opposite direction to that in which the charging current flows

CODE

```
Clc
V=input('Enter the DC Voltage in Volts');
R=input('Enter the Load Resistance in Ohms');
C=input('Enter the Capacitance in Farad');

t1=input('Enter the Range of Charging time in seconds');
t2=input('Enter the Range of Discharging Time in Seconds');

A=V/R;
B=1/(R*C);
I1=A*(1-exp(-B*t1));
I2=A*(-1+exp(B*0.5))*exp(-B*t2);

plot(t1,I1,t2,I2)
xlabel('Time (s)');
ylabel('Current (Ampere)');
title('Current V/S Time');

grid on
table1=[t1' I1'];
table2=[t2' I2'];
table=[table1 table2];
disp('-------------------------------------');
disp('    Table1                Table2');
disp('    Charging            Discharging ');
disp('  Time[t1] Current[I1] Time t2 Current [I2]')
disp('    Sec     Amp !!    Sec      Amp');
disp(table)
disp('-------------------------------------');
disp('----------END OF TABLE -----------');
```

OUPTUT

```
Command Window

File   Edit   Debug   Desktop   Window   Help

New to MATLAB? Watch this Video, see Demos, or read Getting Started.

   Enter the DC Voltage in Volts12
   Enter the Load Resistance in Ohms10
   Enter the Capacitance in Farad3e-2
   Enter the Range of Charging time in secondslinspace(0,0.5,20)
   Enter the Range of Discharging Time in Secondslinspace(0.5,1,20)
   ------------------------------------
      Table1                  Table2
      Charging                Discharging
   Time[t1] Current[I1] Time t2 Current [I2]
      Sec        Amp !!      Sec        Amp
         0          0     0.5000     0.9733
      0.0263     0.1008    0.5263     0.8916
      0.0526     0.1931    0.5526     0.8167
      0.0789     0.2777    0.5789     0.7481
      0.1053     0.3551    0.6053     0.6853
      0.1316     0.4261    0.6316     0.6278
      0.1579     0.4911    0.6579     0.5750
      0.1842     0.5506    0.6842     0.5267
      0.2105     0.6051    0.7105     0.4825
      0.2368     0.6551    0.7368     0.4420
      0.2632     0.7009    0.7632     0.4049
      0.2895     0.7428    0.7895     0.3709
      0.3158     0.7812    0.8158     0.3397
      0.3421     0.8164    0.8421     0.3112
      0.3684     0.8486    0.8684     0.2851
      0.3947     0.8781    0.8947     0.2611
      0.4211     0.9051    0.9211     0.2392
      0.4474     0.9299    0.9474     0.2191
      0.4737     0.9526    0.9737     0.2007
      0.5000     0.9733    1.0000     0.1838

   ------------------------------------
   ----------END OF TABLE -----------
fx >>
```

OUTPUT PLOT FOR CHARGING
DISCHARGING OF CAPACITOR

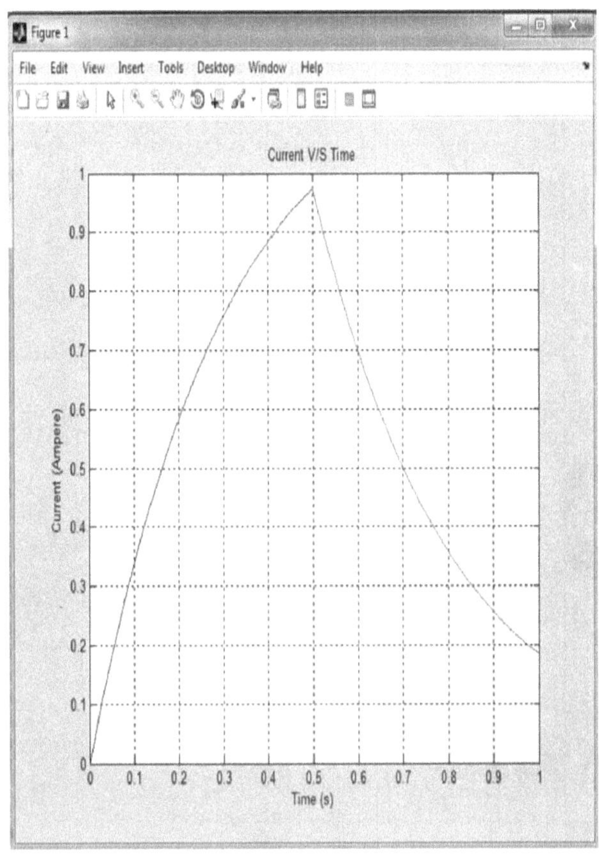

Figure 1
File Edit View Insert Tools Desktop Window Help
Current V/S Time
Current (Ampere)
Time (s)

%Aim:To Determine the Type of Load (RL or RC) and   Plot subgraphs of I vs T,VR vs T,VL vs T,VC vs T

Theory

Transient Response of Series R-L Circuit having DC Excitation

$$V_R = V\left(1 - e^{-\frac{R}{L}t}\right)$$

$$V_L = V e^{-\frac{R}{L}t}, \quad \tau = R/L \text{ is the Time Constant}$$

$$I = V\left(1 - e^{-\frac{R}{L}t}\right)$$

Transient Response of Series R-C Circuit having DC Excitation

$$I = \frac{V}{R} e^{-\frac{t}{RC}}$$

$$V_C = V\left(1 - e^{-\frac{t}{RC}}\right)$$

$$V_R = V e^{-\frac{t}{RC}}$$

Circuit Diagram

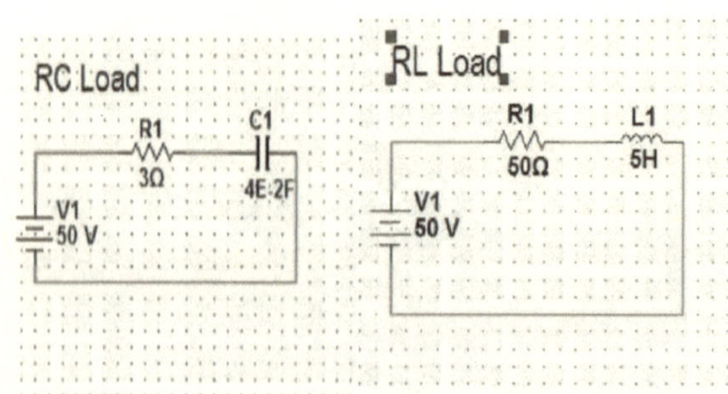

CODE

```matlab
clc
v=input('enter the source voltage in volt=');
disp('enter what kind of load are you using')

disp('for RL load,x=1')
disp('for RC load,x=2')
x=input('enter the value of x=');
if x==1
    disp('your load is Rl')
    R=input('enter the load resistance in ohms=');
    L=input('enter the load inductance in henry=');
    t=input('enter the range of energy storing time=');
     Io=v/R;
     tau=L/R;
     i=Io*(1-exp(-t/tau));
     VR=i*R;
     VL=v*exp(-t/tau);
     Table=[t' i' VR' VL'];
     subplot(3,1,3)
     plot(t,VL)
    grid on
    xlabel('Time(sec)')
    ylabel('VL(volts)')
else if x==2
        disp('your load is RC')
    R=input('enter the load resistance in ohms=');
    C=input('enter the load capacitance in farad=');
    t=input('enter the range of charging time=');
     Io=v/R;
     tau=(R*C);
     i=Io*exp(-t/tau);
     VR=i*R;
     Vc=v*(1-exp(-t/tau));
     Table=[t' i' VR' Vc'];
     subplot(3,1,3)
     plot(t,Vc)
```

```matlab
    grid on
    xlabel('Time(sec)')
    ylabel('Vc(volts)')
    end
end
subplot(3,1,1)
plot(t,i)
grid on
xlabel('Time(sec)')
ylabel('current(ampere)')
subplot(3,1,2)
plot(t,VR)
grid on
xlabel(' Time(s)')
ylabel(' VR(volts)')
disp('------------------------------------------')
disp('!!          Table            !!')
disp('------------------------------------------')
if x==1
   disp('!!  time  !!current!!  VR!!  VL!!')
   disp('!!(sec)   !!(Amp) !! (volt)!! (volt)!!')
else if x==2
     disp('!!  time  !!current!!  VR!!  VC!!')
   disp('!!(sec)   !!(Amp) !! (volt)!! (volt)!!')
   end
end
disp('------------------------------------------')
disp(Table)
disp('------------------------------------------')
disp('!!          END OF TABLE          !!')
disp('------------------------------------------')
```

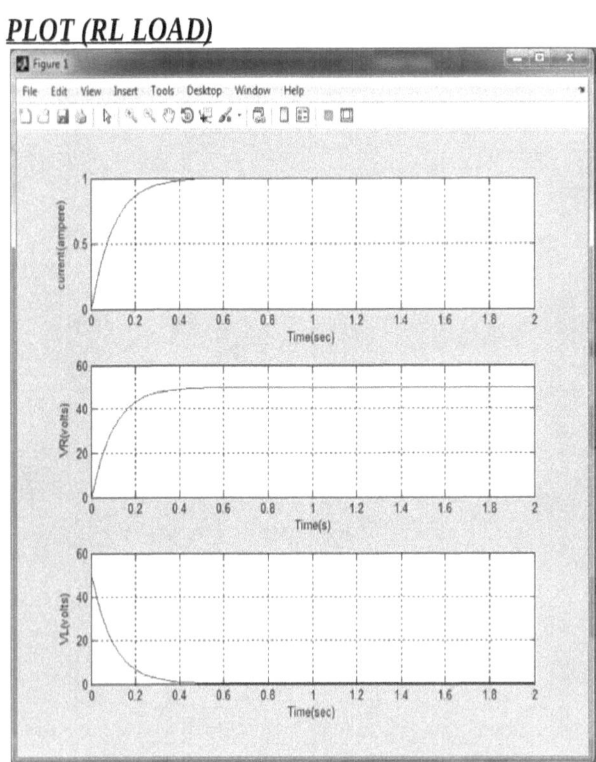

Output RC LOAD

```
Command Window
File  Edit  Debug  Desktop  Window  Help
New to MATLAB? Watch this Video, see Demos, or read Getting Started.
enter the source voltage in volt=50
enter what kind of load are you using
for RL load,x=1
for RC load,x=2
enter the value of x=2
your load is RC
enter the load resistance in ohms=3
enter the load capacitance in farad=4e-2
enter the range of charging time=linspace(0,2,50)
-----------------------------------------
!!            Table                     !!
-----------------------------------------
!!  time  !!current!!   VR!!  VC!!
!!(sec)   !!(Amp)  !! (volt)!! (volt)!!
-------------------------------------------
        0   16.6667   50.0000         0
   0.0408   11.8612   35.5837   14.4163
   0.0816    8.4413   25.3240   24.6760
   0.1224    6.0075   18.0224   31.9776
   0.1633    4.2754   12.8261   37.1739
   0.2041    3.0427    9.1280   40.8720
   0.2449    2.1654    6.4961   43.5039
   0.2857    1.5410    4.6231   45.3769
   0.3265    1.0967    3.2902   46.7098
   0.3673    0.7805    2.3415   47.6585
   0.4082    0.5555    1.6664   48.3336
   0.4490    0.3953    1.1859   48.8141
   0.4898    0.2813    0.8440   49.1560
   0.5306    0.2002    0.6006   49.3994
   0.5714    0.1425    0.4275   49.5725
   0.6122    0.1014    0.3042   49.6958
   0.6531    0.0722    0.2165   49.7835
   0.6939    0.0514    0.1541   49.8459
fx 0.7347    0.0366    0.1097   49.8903
```

Output Waveform (RC LOAD)

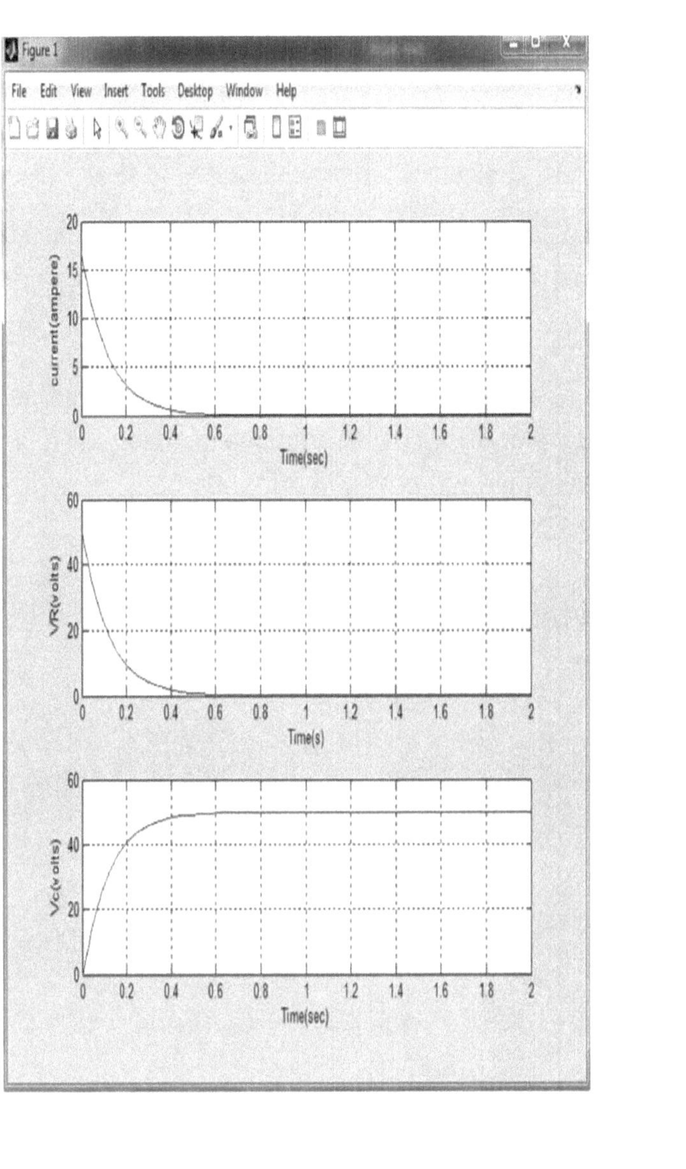

Figure 1
File   Edit   View   Insert   Tools   Desktop   Window   Help
20
15
10
5
0
current(ampere)
0   0.2   0.4   0.6   0.8   1   1.2   1.4   1.6   1.8   2
Time(sec)
60
40
20
0
VR(volts)
0   0.2   0.4   0.6   0.8   1   1.2   1.4   1.6   1.8   2
Time(s)
60
40
20
0
Vc(volts)
0   0.2   0.4   0.6   0.8   1   1.2   1.4   1.6   1.8   2
Time(sec)

Experiment No.10

%Aim: To Determine the Power Efficiency of a Load considering the Impedance of Transmission Lines also

Theory

A transmission line is a system of conductors connecting one point to another and along which electromagnetic energy can be sent. Thus telephone lines and power distribution lines are typical examples of transmission lines

Circuit Diagram

Code

```
clc
Vs=input('enter the value of source voltage=');

 RT=input('enter the value of transmission resistance=');
XT=input('enter the value of transmission Reactance=');
RL=input('enter the value of load resistance=');
XL=input('enter the value of load Reactance=');

ZT=RT+j*XT;
ZL=RL+j*XL;
```

```matlab
Zs=ZT+ZL;
mZs=abs(Zs);
mZL=abs(ZL);
Is=Vs/mZs;
PL=Is^2*RL;
PT=Is^2*RT;
eta=PL/(PT+PL);
fprintf('\n Power Dissipated across Load is %f Watts',PL);
fprintf('\n Power Dissipated Across Transmission Line is %f
watts',PT);
fprintf('\n The Power Efficiency is %f',eta);
```

Experiment No. 11

%Aim: Use of Matlab to Compute the RMS Value of Voltage Waveform for the Given Wave

Theory

The ***Effective or RMS(Root Mean Square***) Value of an alternating current is that steady current (dc) which when flowing through a given resistance for a given time produces the same amount of heat as produced by the alternating current when flowing through the same resistance for the same time

For a Sinusoidal Wave

$$V_{rms}=\sqrt{\frac{V1^2+V2^2+V3^2+\cdots+Vn^2}{n}}$$

The Rms Value of a Symmetrical Wave:

$$V_{rms}=\sqrt{\frac{\textit{Area of half cycle of squared wave}}{\textit{Half Cycle of Base}}}$$

Importance of RMS Values

- The Domestic AC Supply is 230V,50 Hz.It is the RMS Value which means that alternating voltage available has the same heating effect as 230V DC. The Equation is:

$$V = Vm * \sin \omega t = 230 * \sqrt{2} \sin 314t$$

- Ammeters and Voltmeters record r.m.s values of current and voltage respectively

CODE:

```matlab
clc
vm=input('Enter the Peak value of AC Voltage=');
N=input('Enter the range of N=');
n=input('Enter the Value of n for even harmonics');
t=input('Enter the range of time=');
vo=2*vm/pi;
vn=4*vm./(pi*(n.^2-1));
theta=n'*t;
phi=pi*ones(size(theta));
v=vo+vn*cos(theta-phi);
plot(t,v)
xlabel('Time(sec)')
ylabel('Voltage(volts)')
title('Voltage vs Time')
grid on;
Vrms=sqrt(vo^2+0.5*sum(vn.^2));
fprintf('\n Ths RMS Value is %f Volts',Vrms);
```

Output Waveform

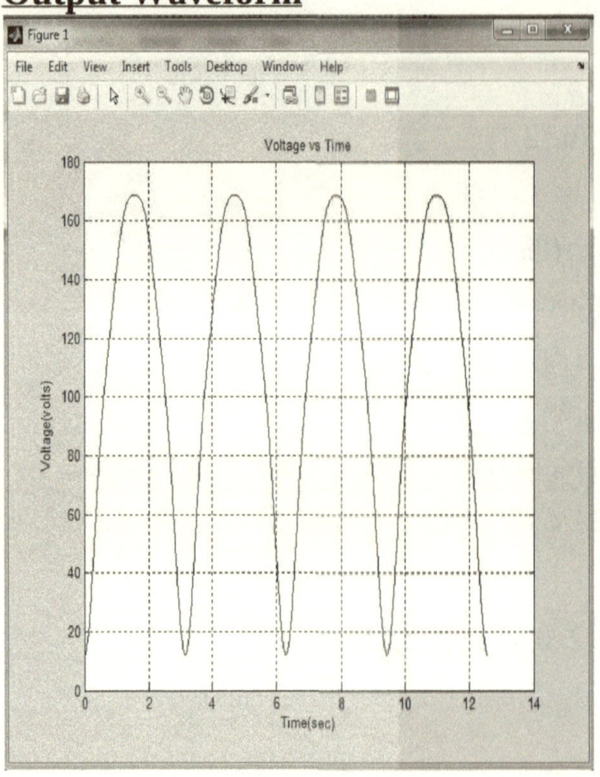

www.ingramcontent.com/pod-product-compliance
Lightning Source LLC
Chambersburg PA
CBHW030417160726
47992CB00007B/3160